# SQL For Beginners

# SQL Made Easy

*A Step-By-Step Guide to SQL Programming for the Beginner, Intermediate and Advanced User (Including Projects and Exercises)*

# Introduction

Structured Query Language or SQL, pronounced as see-Qwell,' is a programming language used to perform operations on a database. These operations include creating, updating, and deleting records. It also allows for the creation and deletion of tables within a database. SQL is just a query language not a database.

To use the SQL query language, you require a database to use it against; there are several types of SQL databases such as MySQL, PostgreSQL, Oracle, SQL Server, etc.

**NOTE:** SQL is not a programming language per se; this is because it lacks features such as loops, methods, objected-orientation, and other core features present in most programming languages. The best way to think of SQL is to view it as a data manipulation language.

Let us define SQL and the essential role it plays in database operations:

# What is SQL

- ❖ SQL is a Structured Query Language

- ❖ We normally use it to manage data in a Relational Database Management System (RDBMS)

- ❖ It is based on relational algebra and relation calculus

SQL plays a very important role in database manipulation. The following are some of SQLs supported operations:

- ❖ Creation of new databases

- ❖ Creating new tables and views

- ❖ Inserting records into a database

- ❖ Updating database records

- ❖ Deleting records from a database

- ❖ Retrieving data from a database

The following are some of the operations that SQL allows users to perform on a database.

- ❖ Allows users to query the data from the database using English-like commands

- ❖ It allows users to describe the data

- ❖ Allows users to set permissions to tables, views, and procedures

- ❖ Allows users to create or drop databases and tables

- ❖ Allows users to create views, procedures, and functions stored in the database

Now that we have seen what SQL is and what it allows us to do, here is a brief history of how it came into being.

PS: Don't forget to leave a review of this book on Amazon if you learn something.

## History of SQL

The SQL programming language is the brainchild of two IBM researchers: Raymond Boyce and Donald Chamberlin who developed it in the early 1970s following a paper on relational models authored by Edgar Todd.

First referred to as SEQUEL, i.e. Structured English Query Language, its primary uses then were to retrieve and perform data operations on the IBM's System/R Relational Database Management System.

In the late 1970s, Oracle Corporation, then known as Relational Software, created their own RDBMS known as Oracle V2. Since then, SQL has advanced and improved significantly and today, we have more than 10 SQL based RDBMS. You can learn more about SQL's history from Wikipedia:

https://bit.ly/1IT9vUm

*SQL For Beginners*

To learn SQL, you need to have a firm understanding of databases. The first sections of this guide introduce you to working with databases:

# Table of Contents

**Introduction** _____ 2

   What is SQL _____ 3

   History of SQL _____ 5

**Section 1: Introduction to Databases** _____ 15

   File-Based Database Models _____ 16

      Disadvantages _____ 17

   Hierarchical Database Model _____ 19

      Advantages _____ 20

      Disadvantages _____ 20

   Network Data Model _____ 21

      Advantages _____ 22

      Disadvantages _____ 22

   Relational Databases _____ 23

      Advantages _____ 24

   Database Schemas _____ 25

   SQL Databases Vs NoSQL Databases _____ 28

## Section 2: Relational Database Management System _____ 32

   SQL Tables _____ 34

   SQL Fields _____ 35

   SQL Records _____ 35

   SQL Columns _____ 36

   NULL Values _____ 37

   SQL Constraints _____ 37

   Data Integrity _____ 39

      Entity Data Integrity _____ 40

      Referential Data Integrity ____ 41

      Domain Data Integrity _____ 42

      User-Defined Data Integrity __ 42

   Normalization _____ 43

      Database Normalization Levels _ 44

   DBMS VS RDBMS _____ 47

## Section 3: Environment Setup _____ 51

   Installing PostgreSQL _____ 52

Installing PgAdmin 4 —————————————— 54

Installing Sample Database ——————————— 56

PostgreSQL Command-Line ———————————— 62

    Adding PostgreSQL to path————————— 62

    Launching PostgreSQL ———————————— 64

    PostgreSQL command line basics ————— 65

## Section 4: SQL Syntax ———————————————— 70

    Essential SQL Commands ———————————— 71

## Section 5: SQL Data Types ——————————— 74

    Text/Character Data Types ———————————— 76

    Numeric Data Types———————————————— 76

    Date and Time Types —————————————— 78

    Network Address Types ————————————— 80

    Geometric Types ————————————————— 81

    Enumerated Data Types ————————————— 81

    UUID Data Types ————————————————— 83

    XML and JSON types ——————————————— 84

## Section 6: SQL Operators — 85

SQL Arithmetic Operators — 85

SQL Comparison Operators — 87

SQL Logical Operators — 90

## Section 7: Working with SQL Databases — 93

How to Create an SQL Database — 93

How to Drop Database in SQL — 95

How to Select a Database in SQL — 97

How to Rename Database in SQL — 99

How to Copy a Database in SQL — 100

## Section 8: Working with SQL Tables — 101

SQL Table Variable — 103

How to Create a Table in SQL Databases — 103

    How to create a table in pgAdmin — 104

How to Use the SQL Drop Table Statement — 108

How to Use the SQL Delete Table Statement — 110

How to Use the SQL Truncate Table Statement 111

Drop vs. Truncate _____ 111

Delete vs Truncate _____ 112

How Use the SQL Rename Table Command __ 113

How to Copy an SQL Table _____ 113

How to Work With SQL Temp Tables _____ 114

   1: Local Temp Table_____ 115

   2: Global Temp Table _____ 115

   PostgreSQL temp Table example _____ 116

How to Use the SQL Alter Table Command __ 117

   1: Adding columns_____ 117

   2: Modifying Columns _____ 118

   3: Renaming Columns _____ 118

   4: Deleting Tables _____ 119

## Section 9: SQL SELECT Query_____ 120

Optional Clause_____ 122

SQL Select Distinct _____ 123

SQL Select Unique _____ 124

SQL Select Count_____ 124

SQL Select Top _____ 126

SQL Select Last ............................................................. 127

SQL Select First ............................................................ 128

SQL Select Random ...................................................... 129

SQL Select Multiple ..................................................... 130

SQL Select Date ............................................................ 130

SQL Select Null ............................................................ 130

## Section 10: SQL Clauses .......................................... 132

SQL - WHERE Clause ................................................. 132

SQL – AND and OR Clause ........................................ 135

SQL – WITH Clause .................................................... 136

## Section 11: SQL ORDER ........................................... 137

Order By ........................................................................ 137

Order By Ascending .................................................... 139

Order by Descending .................................................. 139

Order Data Randomly ................................................ 140

## Section 12: SQL INSERT .......................................... 141

SQL Insert Multiple .................................................... 142

## Section 13: SQL JOIN — 144

- SQL Inner Join — 145
- Right Outer Join — 148
- Left Outer Join — 149
- SQL Full Join — 149
- Cartesian/Cross Join — 150

## Section 14: SQL KEYS — 151

- Primary Keys — 154
  - Primary key rules — 154
- Super Keys — 156
- Candidate Keys — 156
- Alternate Keys — 157
- Foreign Keys — 158
- Compound Keys — 159
- Composite Keys — 159
- Surrogate Keys — 160

## Section 15: SQL Functions — 161

SQL Numeric Functions ............................... 161

SQL String/Text Functions .......................... 163

SQL Date Functions ................................... 165

**Section 16: SQL Injections ........................ 167**

What is SQL Injection? ................................ 167

**Section 17: A Step-by-Step SQL Data Manipulation Exercise .......................... 171**

Exercises & Projects ................................... 171

- Exercise 1 .............................................. 171
- Exercise 2. ............................................. 172
- Exercise 3 .............................................. 172

**BONUS SECTION! An Interactive SQL Project ................................................ 174**

Step 1 ..................................................... 174

Step 2: .................................................... 174

**Conclusion .............................................. 177**

**Check Out My Other Books ..................... 178**

# Section 1: Introduction to Databases

*"Database: the information you lose when your memory crashes."*

**Dave Barry**

In computing, we use the term 'data' to refer to a collection of a distinct, small unit of information. Computers can store data in various formats such as numbers, videos, audios, documents etc.

We can also classify data as information that we can transform into a desirable format for easier transfer and processing. We usually consider data to be interchangeable, which means we can replace or substitute it with another without loss of function or logic.

A database is a collection of related and organized data for easy access and management. We normally organize databases in form of tables, rows, and columns, thus making it easier to find the required information. Databases allow for storage, retrieval, and management of large amounts of data. We use Structured Query Language or SQL to perform operations on databases.

As discussed earlier, we now have several types of SQL databases, and Databases are now available in various models.

In this section, we are going to look at the various database models and discuss how they differ from one another.

## **File-Based Database Models**

A file-based system is a collection of programs used to perform services that allow the users to access information on the system. Each of the programs in a file-based system manages and controls its own set of data. This creates limiting rules that determine the transportation and use of data.

File-Based Database Models require extensive programming and have their ingrained set of advantages and disadvantages such as:

## *Advantages*

- ❖ Maintains Data integrity

- ❖ Maintains Data consistency

- ❖ Increased productivity compared to some models

- ❖ Data independence

- ❖ Improved data backups and recovery compared to older models

- ❖ Improved system performance

## *Disadvantages*

- ❖ Data dependence

- ❖ Data redundancy and duplication

- ❖ Incompatibility with various data formats

- ❖ Non-flexible

- ❖ Data fragmentation

## Hierarchical Database Model

This database model uses tree like structure for data representation where each record contains its own parent data node. This type of model was mainly used for management system such as Information Management System developed by IBM.

This model structure allows one to many as well as one to one relationship between 2 or more types of data. A typical representation of this model appears below:

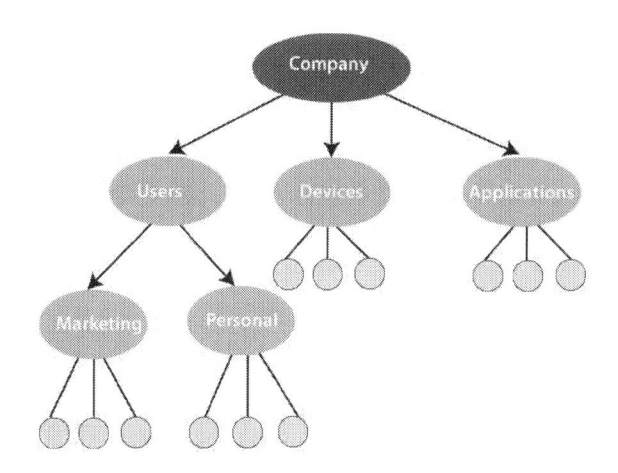

The disadvantages and advantages of this database model are as follows:

## *Advantages*

- ❖ Data retrieval is very fast since this database model provides explicit links between the tables.

- ❖ Built-in and automatically enforced referential integrity

## *Disadvantages*

- ❖ It is sometimes difficult to record in the child table that is unrelated to any record.

- ❖ The created database cannot support complex database relationships.

- ❖ Data redundancy is a very common instance in this model leading to inaccurate information

## Network Data Model

Charles Bachman invented the network model in the early 1960s. In this model, organization of the files in the system is in the form of members and owners. This model is an extension of the hierarchical model, and its primary use is to map many-to-many data relationships.

The network model provides a unique feature in its database scheme, a feature seen as a graph where relationship types are arcs and object types are nodes. This model was the most widely used database model before the introduction of other models such as relational model. The following is typical representation of this model.

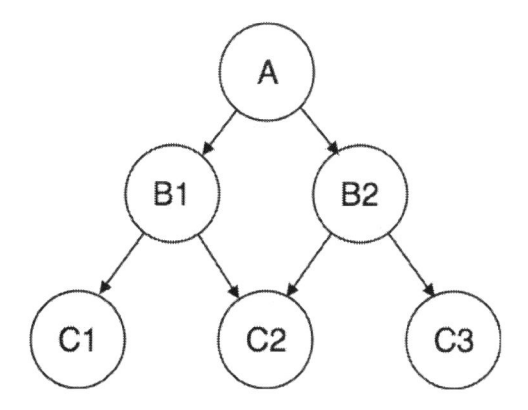

The following are advantages and disadvantages of the network model:

## *Advantages*

- ❖ Simple and easy implementation
- ❖ Supports one-to-one and many-to-many database relationships
- ❖ Provides a great relationship between the child and parent segment
- ❖ Data independence

## *Disadvantages*

- ❖ Functional Errors
- ❖ Provides a complex system structure
- ❖ Structural independence is weak

## Relational Databases

A relational database is the most commonly used database model. In this model, organization of the data is in the form of tables, columns, and records. In a relational database, the defined tables can communicate with each other. This allows for sharing of information thus enhancing features such as searching, organizing, and reporting of data.

As you may have guessed, the Relational Database Model uses the Structured Query Language for programming and user interaction.

The Relational Database Model arranges data in different methods. A relation, which is a term used to refer to each table, contains one or more category columns. We then use a table record or a row to store unique data that we then define to a corresponding column category. This model offers several types of relationships such as:

- One-to-One, which is where one record in a certain table relates to another record in another table

- One-to-Many, which is where one record in a table relates to many other records in another table

- Many-to-One, which is where more than one record in a table relates to one record in a certain table

- Many-to-Many, which is where many records in one table relate to other records in a table

The relational database system offers a range of operations such as SELECT, DROP, DELETE. Later sections of this guide shall discuss these operations.

This type of database model has many advantages over other Database model. For instance:

## *Advantages*

- It is highly scalable because we can add new data without changing the existing records

- ❖ It is highly flexible

- ❖ Compared to other models, it is fast in terms of performance

- ❖ It offers high security when it comes to data sharing

We have many other types of database models that are irrelevant to the scope of this book. They include:

- ❖ Object-Oriented Model

- ❖ NoSQL databases (discussed later)

- ❖ Graph Databases

- ❖ Cloud Databases

## Database Schemas

In a database, a schema, pronounced as "skee-mah," is the detailed structure of a database. It contains the schema objects of the database such as table, view, constraints, keys, columns. Simply put, a database schema is the organization structure of a database.

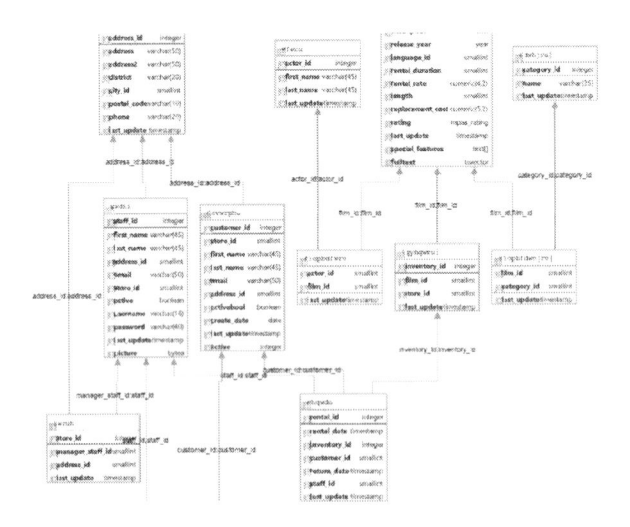

The illustration above shows a schema of a database we are going to use later in the book. You will find tables and view components that make up the database schema.

We can say that Database and schemas are the same or not the same thing. This is because every database engine has its model and design that determines the representation of the database. According to MySQL, a database and a schema are the same. However, in Oracle and SQL Server, the case is different. Therefore, classifying a database as a schema normally depends on the database engine reference.

## SQL Databases Vs NoSQL Databases

Many of the technologies we use today utilize databases. The two most commonly used types of databases are SQL and NoSQL database.

SQL databases are conventional or more traditional types of database where organization of data is in tabular form such as rows and columns. NoSQL databases are new and more efficient database types that do not utilize the tabular model.

In this subsection, we are going to look at the difference between SQL and NoSQL database.

| SQL Databases | NoSQL Databases |
|---|---|
| Use Structured Query Language to manipulate data | Use Unstructured Query Language, which is a set of documents used in data queries |
| SQL databases are scaled vertically | NoSQL are scaled horizontally |
| SQL databases are made up of a fixed schema | Schema in NoSQL database is heavily dynamic for unstructured data. |
| Do not work efficiently with the hierarchical database model | They work perfectly with the hierarchical database model |

| | |
|---|---|
| Support very complex SQL queries | Cannot handle complex queries very well |
| Use tabular-based database representation | Use key-pair values, documents, in-memory graph and search data storage method |
| Include: MySQL, SQLite, Oracle, Microsoft SQL Server, PostgreSQL | Include: MongoDB, Apache Cassandra, CouchDB, Amazon SimpleDB |
| Are relatively Slow | They perform very fast on simple queries compared to SQL databases. |
| Cross-platform and mature | NoSQL databases are new and not fully adopted. |

Now that you have a firmer understanding of databases, we shall build on this knowledge by discussing relational databases a bit further in the next section:

# Section 2: Relational Database Management System

*As you study computer science, you develop this wonderful mental acumen, particularly with relational databases, systems analysis, and artificial intelligence.*

**Frederick Lenz**

A relational database management system is a management system based on the relational database model (discussed in section 1). Its most common representation is as a table that contains rows and columns.

First introduced by E.F Codd, A RDBMS has several key components.

They include:

- Table
- Record/Tuple
- Field/ Column

- ❖ Instance

- ❖ Schema

- ❖ Keys

We will cover some of these components as we progress further into the SQL lessons covered in the book. For now, let us look at tables and columns.

## SQL Tables

Tables are objects used to store data in a Relational Database Management System. Tables contain a group of related data entries with numeric columns and rows.

Tables are the simplest form of data storage for relational databases. They are also a convenient representation of relations. However, a table can contain duplicate rows while a relation cannot contain any duplicate. The following is an example of a DVD rental database table:

| | customer_id [PK] integer | store_id smallint | first_name character varying (45) | last_name character varying (45) |
|---|---|---|---|---|
| 1 | 524 | 1 | Jared | Ely |
| 2 | 1 | 1 | Mary | Smith |
| 3 | 2 | 1 | Patricia | Johnson |
| 4 | 3 | 1 | Linda | Williams |
| 5 | 4 | 2 | Barbara | Jones |
| 6 | 5 | 1 | Elizabeth | Brown |
| 7 | 6 | 2 | Jennifer | Davis |
| 8 | 7 | 1 | Maria | Miller |

The table shows information about customers in a DVD rental database. Later sections of the guide shall focus on retrieving data.

## SQL Fields

Tables break down further into smaller entities referred to as fields. Fields are columns in tables designed to hold information about records in the specified table. For example, in the table above, we can see fields that contain Customer_id, store_id, first_name and last_name.

## SQL Records

In a database, we use Records or Rows to refer to a single entry in a database table. It represents a collection or related data. We can also regard it as the horizontal entity in a database table. For example, in the above table, we have five records or tuples.

| | | | | | |
|---|---|---|---|---|---|
| 1 | 524 | 1 | Jared | | Ely |
| 2 | 1 | 1 | Mary | | Smith |
| 3 | 2 | 1 | Patricia | | Johnson |
| 4 | 3 | 1 | Linda | | Williams |

## SQL Columns

In a database table, we use a column to refer to the vertical entity containing data related to specific fields in a table. For example, in the above table, a column could be first_name or last_name.

| first_name character varying (45) |
|---|
| Jared |
| Mary |
| Patricia |
| Linda |
| Barbara |
| Elizabeth |
| Jennifer |
| Maria |

## NULL Values

A null value is a value assigned to a blank field. This means that a field with a value NULL has no value. In SQL, a zero or a space is not a NULL value. NULL value mainly refers to a field that is blank in a record.

## SQL Constraints

Constraints are rules applied to columns in a database table. We mainly use them to govern the type of data stored within the table. This ensures that the database is accurate and reliable thus minimizing errors.

We can set constraints on columnar or tabular level. Columnar level rules apply only to the specified column while tabular level rules apply to the entire database level.

SQL contains various rules applied to the stored data.

They include:

- **NOT NULL**: This rule ensures that a table or column does not contain null value, which means during record creation, you must enter either a zero, Space, or another value

- **UNIQUE**: Ensures each value is unique and no duplicates are available

- **DEFAULT**: We use this constraint rule to set a default value for a column where we have an unspecified value

- **PRIMARY KEY**: Identifies each record distinctively in a table. We create this rule by combining NOT NULL and UNIQUE constraints

- **FOREIGN KEY:** We use this constraint to identify (distinctively) a record in another database table

- **CHECK** constraint: We use this constraint to confirm if the values in a column fulfill defined conditions

- **INDEX** constraint: We use this to create and retrieve data from a database

# Data Integrity

Data integrity refers to the consistency and accuracy of the data. When creating databases, you MUST pay attention to data integrity. A good database must ensure reinforcement of data integrity as much as possible. Data integrity must also remain maintained during manipulation and update of the database.

Various factors can lead to compromised data integrity within a database. These factors include:

- ❖ Connection failure during transfer of data between databases

- ❖ Data input that is outside the range.

- ❖ Deletion of wrong database records

- ❖ Hacking and malicious attacks

- ❖ Database backup failures

- ❖ Updating of primary key value with the presence of foreign key in a related table

- ❖ Using test data in the database

These primary factors are often the ones that lead to loss of data integrity. To avoid compromised data integrity, it is good practice to back up the database before any operation.

Let us look at the types of data integrity available: These data integrity types exist to each RDBMS:

- ❖ Entity Data Integrity

- ❖ Referential Data Integrity

- ❖ Domain Data Integrity

- ❖ User-Defined Integrity

## *Entity Data Integrity*

Entity data integrity ensures that rows in a database table are unique and thus, no row can be the same within the same database table. The best way to achieve this is by primary key definition. The field in which the primary key is stored contains a unique identifier thus no row can contain the same identifier.

## *Referential Data Integrity*

Referential Data integrity refers to the accuracy and consistency of data within a relationship. A database relationship refers to the data link between two or more tables. In a relationship, we use a certain foreign key to reference another primary key of a certain table. Due to this referencing, we always need to observe data integrity between database relationships.

Hence, Referential Data Integrity requires that the use of a foreign key must reference to an existing and valid primary key.

Referential integrity prevents:

- ❖ Addition of records to a related table if no record is available in the parent table

- Deletion of records in a parent table if records are matching a related table

- Changing values in a primary table resulting in orphaned records in the child table

## *Domain Data Integrity*

Domain data integrity is the consistency and accuracy of data within a column. We achieve this by selecting the suitable data type for a column. Other steps to preserve this type of data integrity could include setting suitable constraints and defining the data forms and range restrictions.

## *User-Defined Data Integrity*

This type of data integrity is custom defined by the database administrator. It allows the administrator to define rules that are not available using any of the other type of data integrity.

# Normalization

Normalization refers to the process of organizing databases to improve data integrity and reduce data redundancy. This process aids the simplification of the database design. Database normalization offers a range of advantages including:

- ❖ Removes the null values available in the databases

- ❖ It helps in query simplification

- ❖ Helps in speed up operations such as searching and sorting the database indexes

- ❖ Helps to clean and optimize database structure

- ❖ Removes data redundancy

- ❖ Helps in achieving compact databases

## *Database Normalization Levels*

Normalization occurs in various levels where each level builds upon the previous levels. Databases must satisfy all the rules of the lower levels to attain a specific level.

Let us discuss the types of database normalization levels. These levels appear arranged in the order of their strength from the strongest to the weakest.

### 1: Domain-Key Normal Form

In this type of normalization level, the relation is to Domain-Key Normal Form (DKNF) when every constraint in the relation follows a logical sequence of the definition keys and domains. This removes the probability of non-temporal anomalies as the domain restraint and enforcing keys roots all the constraints to be met.

### 2: Sixth Normal Form

We say a database relation is in the sixth form normalization when every join dependency of the relation is said to be trivial. We classify a join dependency as trivial if only one of the components is equivalent to the related heading in its total.

## 3: Fifth Normal Form

We say a database relation is in fifth normalization level if non-trivial join dependency in the stated table is indirect by candidate keys.

## 4: Essential Tuple Normal Form

A database relational schema is in ENTF normalization level if it is in Boyce-Codd form and components of every openly declared join dependency are a super key.

## 5: Fourth Normal Form

We say that a database table or relation is in Fourth Normal form if all of its non-trivial multivalued dependencies are super-keys.

**NOTE:** We have many other database normalization levels not covered in this book. They include:

- ❖ Unnormalized Form
- ❖ First Normal Form
- ❖ Second Normal Form
- ❖ Third Normal Form
- ❖ Elementary-Key Normal Form
- ❖ Boyce-Codd Normal Form

You can read more about these types of normalization levels from the resource page below.

https://bit.ly/1NHpdM1

## DBMS VS RDBMS

We have already covered RDBMS. DBMS is not different either. DBMS is a software package we use to create and manipulate databases.

DBMS and RDBMS give programmers and users an organized way of working with databases. DBMS and RDBMS are not very different since both provide a physical database storage.

With the above noted, some relevant difference between them are worth mentioning.

They include:

|   | **RDBMS** | **DBMS** |
|---|---|---|
| 1 | Supports Normalization | Normalization is not available in a DBMS |
| 2 | Supports Distributed Database System | Does not support Distributed Database System |
| 3 | Stores data in a table format | Stores data is a normal computer form |
| 4 | Integrity constraints are defined for the purpose of ACID property | Security for data manipulation is not applied by DBMS |
| 5 | Can handle large amounts of data such | Designed to small scale data and |

| | | |
|---|---|---|
| | as a Company database | personal use |
| 6 | It supports multiple users | Supports a single user |
| 7 | Cases of data redundancy are close to none | Data redundancy is a very common scenario |
| 8 | Stores data in tabular form and utilizes primary keys | Mainly stores data in hierarchical or navigational format |
| 9 | Requires complex software and high-performance hardware to implement | Low software and hardware requirements for implementation |
| 10 | Supports client-server architecture | Client-side architectures is not supported |

| 1 They include PostgreSQL, MySQL and Oracle etc. | Include file system files, xml files, and windows registry. |

At this point in the guide, you know enough to start working with SQL and databases. In the next section, we are going to set up the working environment we shall be using to learn SQL.

# Section 3: Environment Setup

*"In the 21st century, the database is the marketplace."*

**Stan Rapp**

In this section, we are going to look at ways to setup a database system within our local machine. We are going to utilize the RDBMS system. For this book, we are going to use PgAdmin and PostgreSQL.

**NOTE:** Ensure your computer can handle complex database request and operations. The following are the recommended system requirements for the SQL tutorials in this ultimate guidebook:

- ❖ 8GB RAM

- ❖ Intel core i3 and above

- ❖ SSD Storage Device

*SQL For Beginners*

If you are comfortable with using other systems such as Oracle or MySQL, feel free to do so but this book does not cover their setup.

## Installing PostgreSQL

To install PostgreSQL, open your browser and navigate to the following URL.

https://bit.ly/2BDGy2Q

Select your operating system and download the corresponding package. Once you have downloaded, launch the installer, and follow the instruction on the screen.

Setup the installation directory. If you are running Microsoft Windows, select the Program Files folder.

**Installation Directory**

Please specify the directory where PostgreSQL will be installed.

Installation Directory  C:\Program Files\PostgreSQL\11

## SQL For Beginners

On the next prompt, the component's windows, deselect some components such as Stack Builder and PgAdmin as shown below.

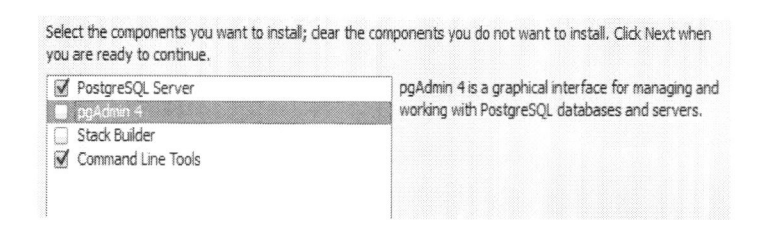

```
Select the components you want to install; clear the components you do not want to install. Click Next when
you are ready to continue.
☑ PostgreSQL Server          pgAdmin 4 is a graphical interface for managing and
■ pgAdmin 4                   working with PostgreSQL databases and servers.
☐ Stack Builder
☑ Command Line Tools
```

Continue following the installation prompts/instructions until you get to the password section. Set up a strong, but easy to remember password. Ensure you remember this password because we are going to use it later.

```
Please provide a password for the database superuser (postgres).
Password          [          ]
Retype password   [          ]
```

Next, setup the port for the PostgreSQL database to connect to; unless you know what you are doing, use the default port.

Please select the port number the server should listen on.

Port 5432

## Installing PgAdmin 4

PgAdmin is an open-source database management tool for PostgreSQL databases. It provides a very convenient interface and working environment suited to beginners and advanced professionals.

In this subsection, we are going to look at how to set it on our computer so that we can use it with PostgreSQL installation we already covered.

Open your browser and navigate to this website to download the latest version of PgAdmin 4.

https://bit.ly/2k4VCDY

**NOTE:** For compatibility purposes, ensure you get the latest version of PgAdmin 4.

*SQL For Beginners*

Once you have downloaded the correct version for your operating system. Launch the launcher and use the installer guide.

Once the installation is complete, click Launch PgAdmin to start the server and to start interacting with the database system. Once you start the server, a browser window will open and prompt you to enter a password. Enter the password we set up using the PostgreSQL installer.

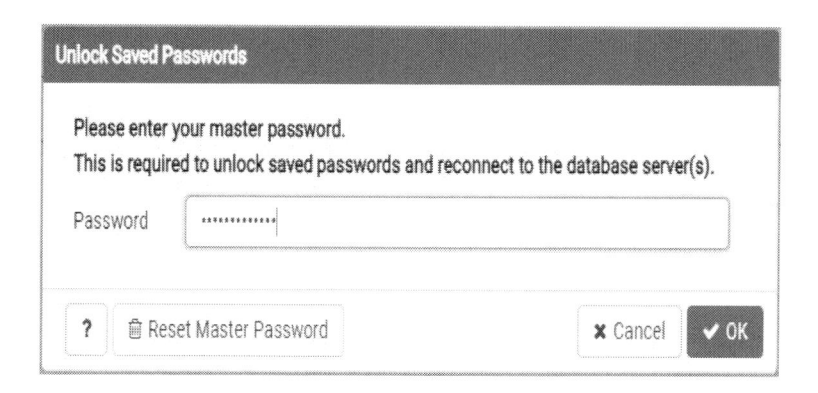

```
Unlock Saved Passwords

Please enter your master password.
This is required to unlock saved passwords and reconnect to the database server(s).

Password     [**************|                              ]

[ ? ]  [ 🗑 Reset Master Password ]              [ ✘ Cancel ] [ ✔ OK ]
```

If the passwords match, a window similar to the one shown below will open up.

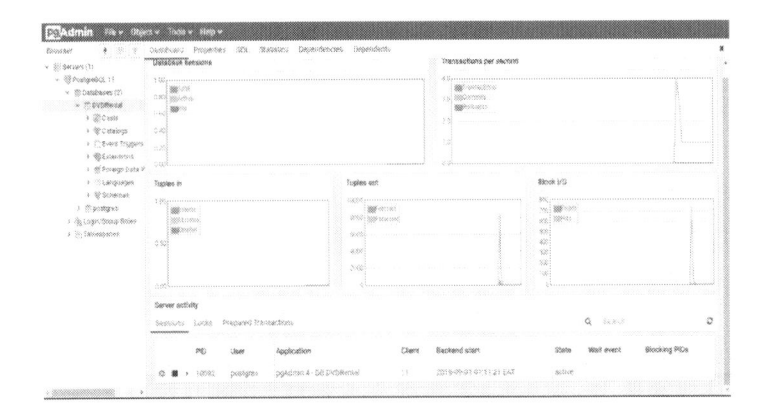

## Installing Sample Database

In this subsection, we are going to look at how to set up a sample database to work with. We are going to use the PostgreSQL sample database available here.

https://bit.ly/2m5IgYZ

Once you have obtained the zip file from the above link, extract it and you will get a dvdrental.tar file. Do not try to extract the file.

Now launch the PostgreSQL browser window. On your left side window, right click the PostgreSQL option and select Create> Database.

*SQL For Beginners*

```
Browser              ⚡ ▦ ▽   Dashboard  Properties  SQL  Statistics  Depend
∨ 🖧 Servers (1)
  ∨  PostgreSQL 11          Database sessions
    ┌─────────────────────┐
    │  Create           ▶ │   Server...
    └─────────────────────┘
       Refresh...            Database...
       Delete/Drop           ┌──────────────────┐
                             │ Login/Group Role... │
       Disconnect Server     └──────────────────┘
                             Tablespace...
       Reload Configuration
       Add Named Restore Point...
       Backup Globals...
       Backup Server...
   >   Properties...           erts
   >   Tablespaces              dates
                                Deletes
```

Next, set the name of the database as DVDRental and click save. Ensure the user is set to postgresql.

*SQL For Beginners*

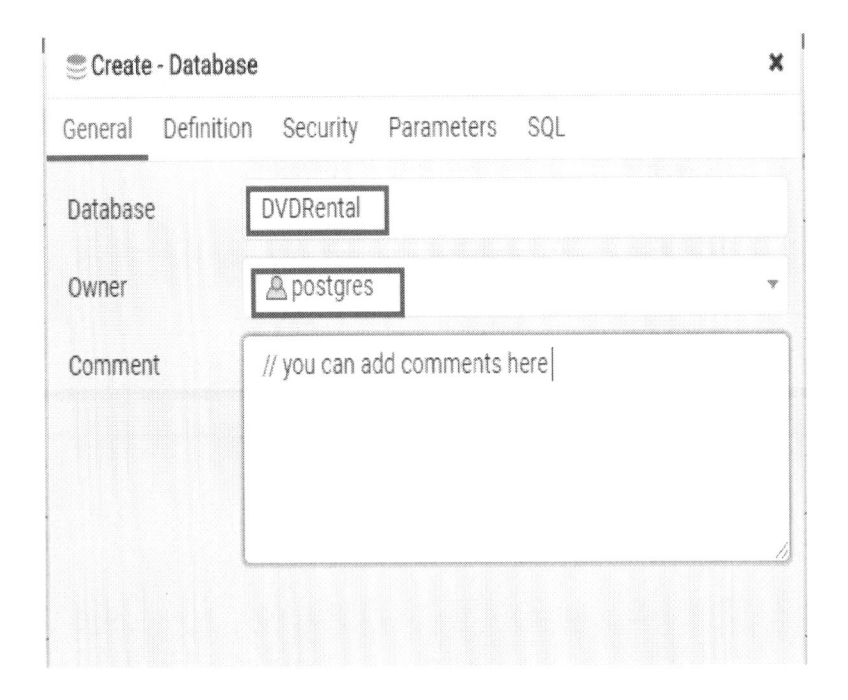

| Create - Database | | | | | ✗ |
|---|---|---|---|---|---|
| General | Definition | Security | Parameters | SQL | |

Database: DVDRental

Owner: postgres

Comment: // you can add comments here

Once the database creation process completes, we can now import the dvdrental.tar sample database we downloaded.

Right click the DVDRental database we just created above and select restore.

Once the restore window appears, select the format as custom or tar. Next, under filename, click the three dots and navigate to the location of the dvdrental.tar file.

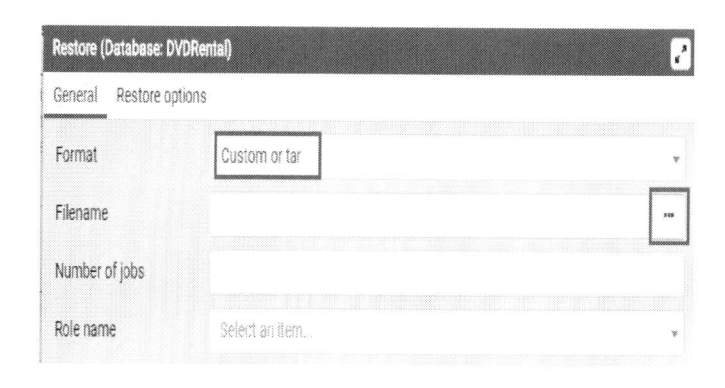

If you cannot see the file, under format, select "all files" and activate show the hidden files. Select the dvdrental.tar file and click OK.

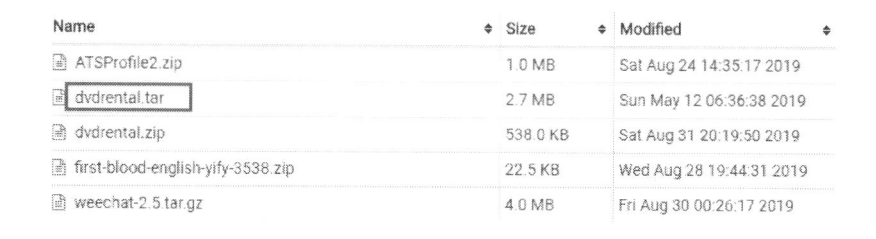

Once you have selected the file, on the same restore window, navigate to restore options and activate the following options: "Pre-Data," "Post-Data" and "Data."

**NOTE:** DO <u>NOT</u> select other options as the database may fail to restore. After setting up the options, select restore to start a restore job on the server.

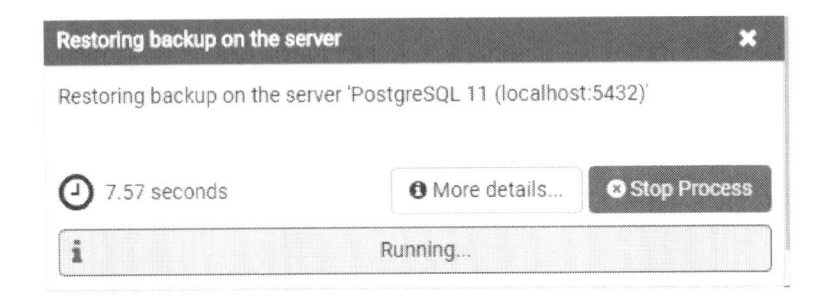

```
Restoring backup on the server                          ×
Restoring backup on the server 'PostgreSQL 11 (localhost:5432)'

  ⏱ 7.57 seconds          ❶ More details...    ⊗ Stop Process
  i                       Running...
```

This database is not too large. Larger databases may take a while based on your system configuration.

Once the database restore has completed, you can start performing SQL queries against it. To test if the database is working correctly, right click the DVDRental database and select open Query Tool.

*SQL For Beginners*

Once the Query window has opened, enter the following SQL query

SELECT * FROM films;

Then press f5 to run the command. This query will display a table with information shown below.

| film_id [PK] integer | title character varying (255) | description text | release_year integer | language_id smallint | rental_duration smallint | rental_rate numeric (4,2) | length smallint | replacement_cost numeric (5,2) |
|---|---|---|---|---|---|---|---|---|
| 1 | 133 Chamber Italian | A Fateful Refl... | 2006 | 1 | 7 | 4.99 | 117 | 14.99 |
| 2 | 384 Grosse Wonderful | A Epic Drama... | 2006 | 1 | 5 | 4.99 | 49 | 19.99 |
| 3 | 8 Airport Pollock | A Epic Tale of... | 2006 | 1 | 6 | 4.99 | 54 | 15.99 |
| 4 | 98 Bright Encounters | A Fateful Yarn... | 2006 | 1 | 4 | 4.99 | 73 | 12.99 |
| 5 | 1 Academy Dinosaur | A Epic Drama... | 2006 | 1 | 6 | 0.99 | 86 | 20.99 |
| 6 | 2 Ace Goldfinger | A Astounding... | 2006 | 1 | 3 | 4.99 | 48 | 12.99 |
| 7 | 3 Adaptation Holes | A Astounding... | 2006 | 1 | 7 | 2.99 | 50 | 18.99 |
| 8 | 4 Affair Prejudice | A Fanciful Do... | 2006 | 1 | 5 | 2.99 | 117 | 26.99 |
| 9 | 5 African Egg | A Fast-Paced... | 2006 | 1 | 6 | 2.99 | 130 | 22.99 |
| 10 | 6 Agent Truman | A Intrepid Pan... | 2006 | 1 | 3 | 2.99 | 169 | 17.99 |

## PostgreSQL Command-Line

Installing PostgreSQL and administering it using PgAdnin is a great way to work with databases. However, when learning SQL, using a graphical control over a database is not ideal. For example, if you are running PgAdmin, to create a database, you just right click and select create new database. The best way to learn SQL is through using the command line.

In this subsection, we are going to look at PostgreSQL database administration tool Psql.

**NOTE:** The command line setup for this section is for Microsoft Windows based machines.

### *Adding PostgreSQL to path*

If you installed the PostgreSQL version specified above, you will not have access to the Psql tool in the command line. To do this in Windows, Open the control panel and navigate to System and Security and then System. Select advanced settings to open a new window, and under the system properties window, select Advanced and Click on Environment Variables.

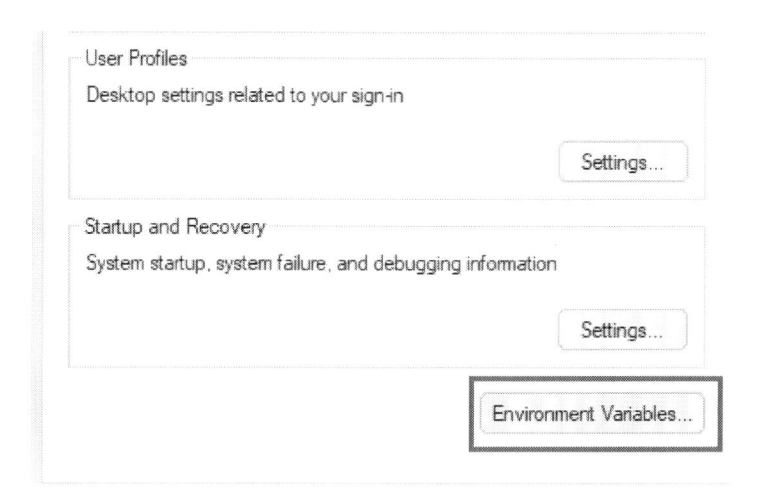

Open the location for the PostgreSQL binary files installation as we specified during installation. Unless you changed it during the installation, the default location is:

C:\Program Files\PostgreSQL\11\bin

Copy the above location. On the new window that opens, navigate to system variables and select path.

| Variable | Value |
|---|---|
| MSMPI_BIN | C:\Program Files\Microsoft MPI\Bin\ |
| NUMBER_OF_PROCESSORS | 4 |
| OS | Windows_NT |
| Path | C:\Program Files\Microsoft MPI\Bin\;C:\ProgramData\Anaconda3;... |
| PATHEXT | .COM;.EXE;.BAT;.CMD;.VBS;.VBE;.JS;.JSE;.WSF;.WSH;.MSC |
| PROCESSOR_ARCHITECTURE | AMD64 |
| PROCESSOR_IDENTIFIER | Intel64 Family 6 Model 42 Stepping 7, GenuineIntel |

Double click the path variable and click add new. Add the Path you copied above and Click OK and close the windows.

## *Launching PostgreSQL*

Once the above operation has completed, open you command window and enter the command psql –version to check if it is working properly. You should get an output as psql (PostgreSQL) 11.5

To connect to the server now, enter the command psql -U postgres. Now enter the password you set up during the installation process. Once connected, you can start performing SQL commands against the connected database.

## PostgreSQL command line basics

The first commands we are going to look at are how to connect with and work with the database.

### 1: Connect database

To connect to a PostgreSQL database, enter the commend \c <database name>. For example, to connect to the DVDRental database we created earlier, \c DVDRental;

### 2: View Databases

To view the database in the server, you use the command \l.

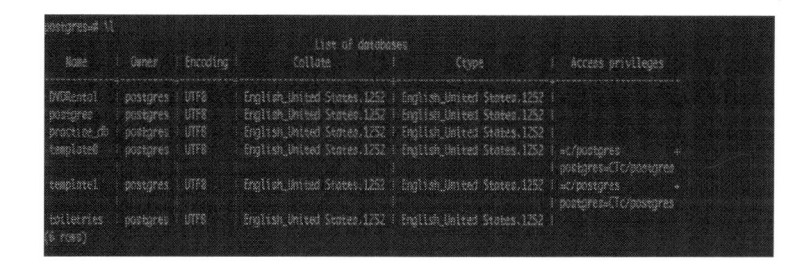

### 3: List tables

You can view the tables in the connected database schema by using the command: \dt

## 4: Describe table

You can connect to a specific table within the connected database by using the command. \d <table_name>;

## : Previous command

You can view the previously executed command by using the \g command.

## 6: Get version

To get the more details about the version of PostgreSQL installed, you can use the **select version()** command.

## 7: Help

If you want to get more information about an SQL command, you can use the \h <SQL command>;

```
DVDRental=# \h CREATE DATABASE;
Command:     CREATE DATABASE
Description: create a new database
Syntax:
CREATE DATABASE name
    [ [ WITH ] [ OWNER [=] user_name ]
           [ TEMPLATE [=] template ]
           [ ENCODING [=] encoding ]
           [ LC_COLLATE [=] lc_collate ]
           [ LC_CTYPE [=] lc_ctype ]
           [ TABLESPACE [=] tablespace_name ]
           [ ALLOW_CONNECTIONS [=] allowconn ]
           [ CONNECTION LIMIT [=] connlimit ]
           [ IS_TEMPLATE [=] istemplate ] ]
```

## 8: Query time

You can also find the time taken to execute an SQL query by using the \timing command.

## 9: List commands

To view all the commands supported by the psql application, you can use the \? to view this commands.

## 10: Text editor

You can interact with the previously executed command in text editor by adding or removing SQL queries. You can do this using the \e command.

Those are the most important commands you need to know to start interacting with PostgreSQL using the command line.

Now that you know this, let us move on to learning more about the SQL syntax

# Section 4: SQL Syntax

*"Always code as if the guy who ends up maintaining your code will be a violent psychopath who knows where you live."*

**Martin Golding**

SQL is very similar to most programming languages in the sense that it too has a set of rules and guidelines that allow for database programming and queries. In this section, we are going to cover the most basic SQL commands. The following are factors to note about SQL:

- ❖ SQL is case insensitive and thus, the commands select and SELECT mean the same

- ❖ You can place SQL statements on one or multiple lines

- ❖ You <u>MUST</u> end SQL statements with a semi-colon.

- ❖ SQL statements rely on relational algebra and tuple calculus

If you are working with MySQL, note that MySQL makes differences in table names thus, when performing queries on a table, provide the exact table name.

The SQL statement syntax is SELECT column_name FROM table_name;

## Essential SQL Commands

In this subsection, we are going to mention most of them but keep in mind that because this is a hands-on SQL guidebook, we may reference them severally in the book. The most essential SQL commands are:

- ❖ **SELECT** – We use this command to extract the specified data from a database

- ❖ **FROM** – We use this command to specify where data extraction in a database is to take place

- ❖ **DELETE** – We use this command to delete the specified data within a database

- **UPDATE** – This command updates data and records in the database

- **CREATE DATABASE** – We us this command to create a new database with a specified name and user

- **ALTER DATABASE** – We use this command to alter or modify an existing database

- **CREATE TABLE** – This command creates a new table within a database

- **DROP TABLE** – We use this command to delete a specified table within a database

- **CREATE INDEX** – We use this command to create a database index or a search key

- **DROP INDEX** – We use this command to delete a specified index within a database

- **INSERT INTO** – This is the command we use when we want to add new data within a database

- ❖ **TRUNCATE TABALE** – We use this command when our aim is to truncate a specified table within a database

SQL has many other commands that we will refer to later in the book.

Now that you know the most important SQL commands, the next step in your learning process is to learn about SQL data types. The next section introduces you to this:

# Section 5: SQL Data Types

*"Torture the data, and it will confess to anything."*

**Ronald Coase**

If you are not new to programming in any other computer programming language, the aspect of data type is familiar. SQL is no different; we can consider an SQL data type as an attribute that determines the type of data to be stored within a specific location.

In an SQL database, component such as Expressions, variables and column have a data type. We normally specify these data types when creating a database.

All SQL databases have three main data types:

- ❖ Numeric data types

- ❖ String Data types

- ❖ Date and Time data types

Data types may vary from one database to another. Since our lessons in this book are using PostgreSQL, we are going to cover all the data types supported by PostgreSQL. They include:

❖ Text Data Types

❖ Numeric Data types

❖ Date and time data types

❖ Network Address data types

❖ Geometric data types

❖ Enumerated data types

❖ UUID types

❖ XML and JSON types

Let us discuss the above data types.

## Text/Character Data Types

In PostgreSQL, the Text data types store text characters. All character data types in PostgreSQL have the same internal structure. The supported text types are:

| varchar(n) | **Declares a variable-length with limit** |
| --- | --- |
| text | Used in variables with unlimited length |
| char | Variables with fixed-length |

## Numeric Data Types

PostgreSQL databases support two main types of numeric data types: Integers and Floating-point numbers.

However, their classification depends on their size limit as follows:

| Data type | Storage size | Range Value |
|---|---|---|
| real | 4 bytes | 6-point precision float |
| double | 8 bytes | 15-point precision float |
| integer | 4 bytes | -2147483648 to 2147483647 |
| bigint | 8 bytes | -9223372036854775808 to 9223372036854775807 |
| smallint | 2 bytes | -32768 to 32767 |
| decimal | Size may | 131072 to 16383 |

|   | vary |   |
|---|------|---|

## Date and Time Types

The Date and time types allow you to store date or time data.

We also refer to them as temporal data types.

They include:

| Type | Size | Range |
|---|---|---|
| date | 4 bytes | 4713 BC to 5874897 AD. |
| Interval | 12 bytes | -178,000,000 years to 178,000,000 years |
| Timez | 12 bytes | 00:00:00 + 1459 to 24:00:00-1459 |
| Time | 8 bytes | 00:00:00 to 24:00:00 |
| Timestampz | 8 bytes | 4713 BC to 294276 AD |
| Timestamp | 8 bytes | 4713 BC to 294276 AD |

You can learn more about date and time type from the following invaluable resource page:

https://bit.ly/2m16Dqp

## Network Address Types

We use network data types to store network information such as IP addresses. If an application has to store network information, PostgreSQL provides three types of network address options.

They are:

- ❖ macaddr – We use this data type to store mac addresses of devices. It has a storage size of 6 bytes.

- ❖ Inet – This data type stores Internet Protocol version 4 and 5 (IPv4 and IPv5) hosts and networks; it has a size of 18 bytes.

- ❖ Cider – This data type stores Internet protocol version 4 and 6 networks; it also has a size of 18 bytes.

## Geometric Types

Geometric data types represent two-dimensional spatial objects. Geometric data types help in geometrical procedures such as rotation, translation, etc.

They include:

- ❖ Point
- ❖ Line
- ❖ Polygon
- ❖ Circle
- ❖ Path
- ❖ Box

## Enumerated Data Types

We use enumerated data types to represent constant information that hardly changes. Examples of constant information include country code, addresses, etc. In a database table, we represent this type of information using foreign keys—to ensure data integrity.

## UUID Data Types

UUID or Universally Unique Identifier is a 128-bit unique number generated using a specific algorithm used to identify information in computers. It mainly comprises of a hexadecimal digit separated using dashes.

In PostgreSQL, we use the UUID data types to store this type of numbers. GUID, or Global Universally Unique Identifier, which are UUID for Global identification, are also stored as a UUID data type. To learn more about UUID, refer to the following resource page:

https://bit.ly/1kw0IGG

The UUID data type takes at least 16 bytes of storage. Here are examples:

- ❖ 40e6215d-b5c8-4896-987c-f30f3678f608
- ❖ 6ecd8c99-4036-403d-bf84-cf8400f67836
- ❖ 8f333df6-90a4-4fda-8dd3-9485d27cee36

## XML and JSON types

We use these two data types to store XML and JSON data respectively. They are not that much different from the text data type except that PostgreSQL checks if the stored data has the proper formatting. JSON data types extend to JSON and JSONB.

# Section 6: SQL Operators

*"Measuring programming progress by lines of code is like measuring aircraft building progress by weight."*

**Bill Gates**

SQL operators are keywords or characters used with the SQL statement to perform specified operations such as comparison and arithmetic operations.

SQL supports three main types of operators namely:

- ❖ Arithmetic operators
- ❖ Comparison Operators
- ❖ Logical Operators

## SQL Arithmetic Operators

We use SQL arithmetic operators to perform mathematical or arithmetic operations. For instance, assume we have variables "x" and "y" where the respective values are 500 and 1000.

In such a case, the following operators would apply:

| Operator | Operation | Example |
|---|---|---|
| + | Addition: adds two operands on both sides | x + y = 1500 |
| - | Subtraction: returns the difference between right hand and left-hand operands | x – y = -500 |
| * | Multiplication: returns the multiple of the provided operands | x * y = 500000 |

| | | |
|---|---|---|
| / | Division: returns the quotient of the provided operands | y / x = 2 |
| % | Modulo: returns the remainder after a division | y mod x = 0 |

## SQL Comparison Operators

Comparison operators are types of operators used to compare two values. In programming, these types of operators return a TRUE or FALSE value. We normally use them in conditions to compare expressions.

The table below shows the SQL comparison operators:

| Operator | Operation |
|---|---|
| = | Equality: used to evaluate if the operands or expressions are equal |
| < | Less Than: evaluates is the left value is less than the right value |
| > | Greater than: evaluates if the left value is greater than the left value |
| <> | Not Equal to: evaluates if the values are not equal |
| !> | evaluates if the left value is not greater than the right value |
| != | Evaluates if the values are |

|  | equal |
| --- | --- |
| >= | Greater than or equal to |
| <= | Less than or equal to |
| !< | Not less than |

## SQL Logical Operators

We use Logical operators to check for logical conditions between expressions; we normally use them with other True or False Operators. SQL supports the

Logical Operators shown in the table below:

| Operator | Operation |
| --- | --- |
| **AND** | Compares Boolean expressions and returns true if all expressions are true |
| **OR** | Compares Boolean expressions and returns true if either one of them is true |
| **NOT** | Returns the reverse of a logical expression |
| **BETWEEN** | Searches for values within a specified |

|  |  |
|---|---|
|  | range |
| **IN** | Compares values within a specified range |
| **ALL** | Compares values with other values in another set of values |
| **ANY** | Compares values in a list of values according to a specified condition |
| **SOME** | Compares a specified value to each value list |
| **EXISTS** | Checks for existence of a row in a table |
| **IS NULL** | Compares a specified value with a NULL value |

*SQL For Beginners*

| | |
|---|---|
| **LIKE** | Compares a value by using wildcard operators such as *, ?, #, [] |

Now that we have covered the logic behind databases, it is time to learn how to work with them.

# Section 7: Working with SQL Databases

*"You can have data without information, but you cannot have information without data."*

*Daniel Keys Moran*

In this section, we are going to learn how to use SQL statements to create, rename, and delete databases. We will focus on SQL commands that help us achieve these tasks.

Let us get started.

## How to Create an SQL Database

To create an SQL database, we use the CREATE DATABASE statement. The basic syntax for creating an SQL database is as follows:

CREATE DATABASE <database nam>;

This statement applies to almost any number of SQL database engines such as PostgreSQL, MySQL, and SQL server. In PostgreSQL, you can use the pgAdmin or the Command-Line to create new databases.

You can also include other parameters in the database creation command. This helps you modify the creation of the database as well as how it works.

Some of the parameters you can include are:

- ❖ role_name – We use this parameter to define the role of the user who created the database.

- ❖ max_concurrent_connection – We use this parameter to specify the number of concurrent connections a database can allow. PostgreSQL allocated—1 by default to set it to unlimited.

- ❖ Encoding – We use this parameter to set the character encoding used in the creation of the database—UTF8 by default.

- ❖ Template – This parameter allows you to specify the template name that you want to create your database.

```
CREATE DATABASE "DemoDatabase"
    WITH
    OWNER = postgres
    ENCODING = 'UTF8'
    LC_COLLATE = 'English_United States.1252'
    LC_CTYPE = 'English_United States.1252'
    TABLESPACE = pg_default
    CONNECTION LIMIT = -1;
```

## How to Drop Database in SQL

We use this SQL statement to remove all the data directories and catalogs in PostgreSQL. This statement performs these actions permanently and you should therefore use it with caution.

The general syntax for database drop is:

DROP DATABASE <database name>;

You can also add parameters to modify the execution of the above command.

The parameters commonly used here include:

- ❖ If exists – We use this parameter to display a warning if the specified database does not exists.

- ❖ -W – We use this parameter to ask for a password when dropping the database.

Other parameters allow you to define how to execute this command.

To drop a database in pgAdmin, right click the database you want to remove and select delete/drop.

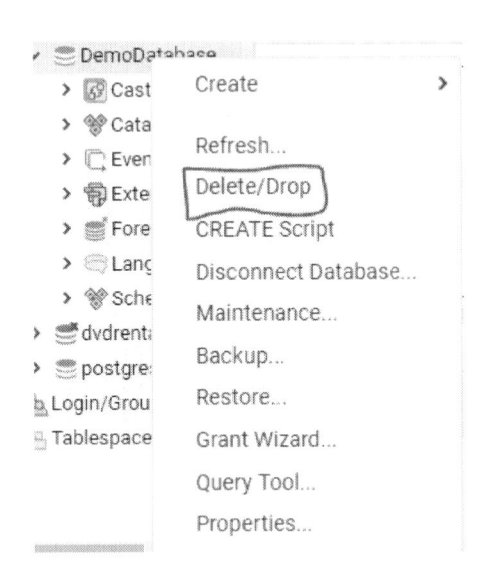

**NOTE:** If you use the drop command against a database, it will permanently delete all the tables, records, and views so be careful when using this command.

## How to Select a Database in SQL

We use this command to select the database you want to work on. If you are using a graphical control such as PgAdmin for PostgreSQL or PhpMyAdmin for MySQL, you will rarely use this command because you can just click to perform these operations. However, for advanced database administration, a command-line interface is the best choice.

To select a database in the command-line, you can use the command USE <database name>; for MySQL, and \c <database name> for PostgreSQL.

```
postgres=# \c DVDRental;
WARNING: Console code page (437) differs from Windows code page (1252)
         8-bit characters might not work correctly. See psql reference
         page "Notes for Windows users" for details.
You are now connected to database "DVDRental" as user "postgres".
DVDRental=#
```

The connection syntax may vary based on the database engine you are using. For commands on the engine command, refer to the help option on the database you are using.

# How to Rename Database in SQL

We use the SQL RENAME statement to change the name of the specified database. Renaming a database may occur if the existing name is irrelevant or if you need to create a database with a close name.

Similar to other SQL statements, this command may vary based on the used engine.

The syntax for rename statement is:

RENAME DATABASE <previous name> TO <new name>;

In PostgreSQL, to rename a database, first disconnect from the database you wish to rename and connect to a new database and then enter the command ALTER DATABASE <previous name> RENAME TO <new name>

```
DVDRental=# \c postgres
WARNING: Console code page (437) differs from Windows code page (1252)
         8-bit characters might not work correctly. See psql reference
         page "Notes for Windows users" for details.
You are now connected to database "postgres" as user "postgres".
postgres=# ALTER DATABASE DVDRental RENAME TO DVDRentalRename;
ALTER DATABASE
postgres=#
```

## How to Copy a Database in SQL

If you want to copy a database within the same server, you can use the template argument in the create database. The syntax may vary based on the engine used. For PostgreSQL, we use the syntax CREATE DATABASE <db name> WITH TEMPLATE <existing database to copy>

```
postgres=# CREATE DATABASE CopiedDB WITH TEMPLATE DVDRental;
CREATE DATABASE
postgres=#
```

There are other cases when restoring databases on different servers. You can copy databases by first dumping the source file. Next, create the database on the target server and then importing the dump file from the source server.

# Section 8: Working with SQL Tables

*"The best programs are written so that computing machines can perform them quickly and so that human beings can understand them clearly. A programmer is ideally an essayist who works with traditional aesthetic and literary forms as well as mathematical concepts, to communicate the way that an algorithm works and to convince a reader that the results will be correct."*

**Donald E. Knuth**

We can classify a database table as a collection of related data ordered well in rows and columns. In databases, we refer to a table as a relation and a row as tuple.

Tables are forms of storage locations.

Let us look at an example of a table using the dvdrental database.

| | country_id<br>[PK] integer | country<br>character varying (50) | last_update<br>timestamp without time zone |
|---|---|---|---|
| 1 | 1 | Afghanistan | 2006-02-15 09:44:00 |
| 2 | 2 | Algeria | 2006-02-15 09:44:00 |
| 3 | 3 | American Samoa | 2006-02-15 09:44:00 |
| 4 | 4 | Angola | 2006-02-15 09:44:00 |
| 5 | 5 | Anguilla | 2006-02-15 09:44:00 |

We generated the table above from the customer table (in the dvdrental database). We can see the columns such as county_id and last_update. The rows are 1, 2, 3... (command used is SELECT * FROM county;)

## SQL Table Variable

Microsoft introduced the SQL table variable for SQL server to act as an alternate for temporary tables. We use it to perform operations such as creating, modifying, renaming, and deleting tables. It acts as a temporary variable to store results and database records and is therefore similar to temp table.

The syntax for table variables is similar to the create table command. However, we cannot use a table variable for input and output parameters.

## How to Create a Table in SQL Databases

To create a table in SQL databases, we use the command CREATE TABLE. During the creation of database tables, we have to define the table name, column, and specific column data type.

Below is a simple syntax for creating an SQL table:

```
CREATE TABLE "table_name"

("column_name" "data_type",

"column_name" "data_type",

"column_name" "data_type",

...

"columnN" "data_type");
```

As discussed in the data type section, the data type of columns may vary based on the engine used. For example, in MySQL and PostgreSQL, we use the INT data type for numbers; on the other hand, in Oracle databases, we use the NUMBER data type.

Let us look at how to create a table in pgAdmin.

### *How to create a table in pgAdmin*

Open the database in which you want to create the table. Click on the schema and right click tables.

Select create new table:

Enter the table name and click on columns in the top menu. Click on the plus icon to add a column. Set the column name and the data type. If you want to set the column as a primary key, click set it as primary key.

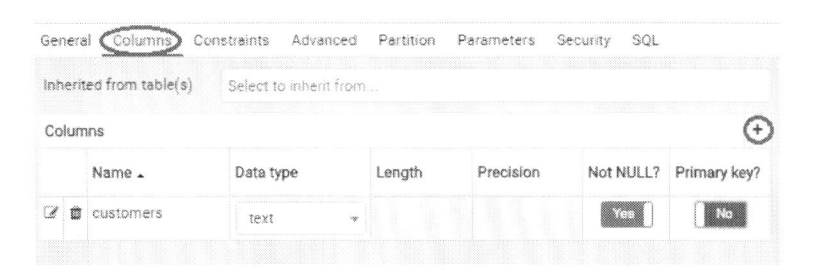

This is a very easy way of working with PostgreSQL databases. To create a table in the command-line, use the command discussed above. For example,

```
postgres=# CREATE TABLE CUSTOMERS(
postgres(#    ID      INT               NOT NULL,
postgres(#    NAME    VARCHAR (20)      NOT NULL,
postgres(#    AGE     INT               NOT NULL,
postgres(#    ADDRESS CHAR (25) ,
postgres(#    SALARY  DECIMAL (18, 2),
postgres(#    PRIMARY KEY (ID)
postgres(# );
CREATE TABLE
```

The message CREATE TABLE means that we have successfully created a table. If it exists, PostgreSQL will report an error.

```
postgres=# CREATE TABLE CUSTOMERS(
postgres(#    ID      INT               NOT NULL,
postgres(#    NAME    VARCHAR (20)      NOT NULL,
postgres(#    AGE     INT               NOT NULL,
postgres(#    ADDRESS CHAR (25) ,
postgres(#    SALARY  DECIMAL (18, 2),
postgres(#    PRIMARY KEY (ID)
postgres(# );
ERROR:  relation "customers" already exists
postgres=#
```

To view the information stored in the table, you can use the DESC command for MySQL or \d <table name> for PostgreSQL. As shown below:

```
postgres=# \d customers
                  Table "public.customers"
 Column  |         Type          | Collation | Nullable | Default
---------+-----------------------+-----------+----------+---------
 id      | integer               |           | not null |
 name    | character varying(20) |           | not null |
 age     | integer               |           | not null |
 address | character(25)         |           |          |
 salary  | numeric(18,2)         |           |          |
Indexes:
    "customers_pkey" PRIMARY KEY, btree (id)

postgres=#
```

**NOTE:** Creating SQL tables may vary from one engine to another.

## How to Use the SQL Drop Table Statement

We use the SQL DROP TABLE statement to delete/remove database table indexes. If you have a database within another database schema, you can use the DROP DATABASE instead. When used, this command permanently clears all the database table data, triggers, permissions, and constraints. This command is similar to the SQL DROP DATABASE.

**NOTE:** This command clears data recursively thus no probability of data recovery. The basic syntax for the database table drop command is:

DROP TABLE <name>;

As is the case with other statements, you can use other parameters to control the deletion of the table. These arguments include:

- ❖ IF EXISTS – We use this parameter to return a warning if the table does not exist

❖ RESTRICT – This parameter prevents table deletion if there are objects dependent on the table

❖ CASCADE – We use this parameter to delete all the objects that are dependent on the specified table

For illusory purposes, we shall now use the drop statement to delete the customers table we created earlier. Connect to SQL using command prompt or launch the Query Tool in pgAdmin.

```
postgres=# DROP TABLE customers;
DROP TABLE
postgres=# \d customers
Did not find any relation named "customers".
postgres=#
```

The execution of the same command on pgAdmin is similar.

```
DROP TABLE customers;
```

## How to Use the SQL Delete Table Statement

We use the SQL delete statement together with the statement WHERE to delete specific rows in a table. If you want to delete all rows in a table, you do not include the WHERE condition. The SQL statements DELETE, TRUNCATE (discussed below), and DROP may appear similar but they have exceptional differences.

The syntax for table row deletion is:

```
DELETE FROM <table_name> [WHERE];
```

You can also use the DELETE FROM <table_name>; to delete all the rows in the specified table.

# How to Use the SQL Truncate Table Statement

We use the SQL TRUNCATE statement to remove all the rows from the specified table. This statement mainly acts as the DELETE statement without the WHERE condition. The syntax for table truncation is:

```
TRUNCATE TABLE <table_name>;
```

## *Drop vs. Truncate*

We use the SQL drop statement to drop the table rows and its corresponding definition; this fully destroys the relationships between the table and other tables with the database.

The following are some of the operations the drop commands executes:

- ❖ Completely removes all the relationships

- ❖ Completely purges user privileges

- ❖ Removes Data integrity constraints

❖ Removes Entire table structure

On the other hand, the TRUNCATE command does not remove the table structure and some issues described above may not be available.

## *Delete vs Truncate*

The differences between these two commands are dismal. The DELETE command deletes the rows in the specified table. However, it does not remove the space occupied by the table. The truncate command removes all the rows within the table and the space they previously occupied.

## How Use the SQL Rename Table Command

Similar to the RENAME DATABASE command, we use the RENAME TABLE command to change the names of the specified table. Similar cases to rename may occur as those of renaming databases. The basic syntax for table renaming is:

```
ALTER    TABLE    <table_name>    RENAME    TO
<new_table_name>;
```

## How to Copy an SQL Table

A case may rise where we need to copy a table to another table in the same database. We can use the SELECT and INTO commands to accomplish this task. The syntax for table copying is as follows:

```
SELECT * INTO <table_to_copy_to> FROM
<table_to_copy_from> WHERE condition;
```

## How to Work With SQL Temp Tables

Let us discuss the concept of temp tables first introduced in the SQL Server and then adopted by other engines. Temp, or temporally tables, are created during run-time. Temp tables perform operations similar to other tables within the tempdb database.

As the name suggests, Temp tables have a limited lifetime and their deletion usually occurs once the database session has ended. In addition, depending on the type of temp tables created, a temp table can be visible on the session that created it or other session may be able to access it.

We normally have two main types of database temp tables created based on the scope. They include:

- ❖ Local Temp table

- ❖ Global Temp table

## 1: *Local Temp Table*

Normally, Local temp tables have limited scope and are only available to the session used to create them. They automatically 'get dropped' once you close a database session. We mainly create them using the # sign. The syntax for local temp table creation is:

```
CREATE TABLE #local temp table;
```

## 2: *Global Temp Table*

On the other hand, global temp tables are similar to permanent tables and are thus accessible on all sessions of the database unless deleted. We mainly create them using two ## hash signs. The general syntax for creating a global temp table is:

```
CREATE TABLE ##global temp table;
```

As we have seen, the temp table creation command may vary on the engine used.

## *PostgreSQL temp Table example*

We shall now look at an example of how to create a temp table in PostgreSQL. Connect to the PostgreSQL server and open the Query Tool (pgAdmin) or use the command-line.

First, create a database called temp and connect to it.

Use the following commands to create a temp table called temp_table

```
CREATE TEMP TABLE temp_table(c INT);
```

Then select all the rows within the table

```
SELECT * FROM temp_table;
```

To test whether the table is accessible to other sessions, open a new connection to the test database and enter the command SELECT * FROM temp_table;

```
postgres=# \c temp
WARNING: Console code page (437) differs from Windows code page (1252)
         8-bit characters might not work correctly. See psql reference
         page "Notes for Windows users" for details.
You are now connected to database "temp" as user "postgres".
temp=# SELECT * FROM temp_table;
ERROR:  relation "temp_table" does not exist
LINE 1: SELECT * FROM temp_table;
                      ^
temp=#
```

You can see that trying to use the temp_table returns an error indicating that it does not exist. To drop the temp table, you can use the DROP TABLE command.

## How to Use the SQL Alter Table Command

We usually use the SQL ALTER command to modify, delete, or add columns to a table. You may also use it to rename tables, and to add and delete table constraints.

### 1: Adding columns

If you want to add columns to an existing table, you can use the SQL ALTER TABLE command using the syntax below.

```
ALTER TABLE <table_name>

ADD (<column1> <column1-definition,

<column2> <column2-definition,

<columnN> <columnN-definition>);
```

## 2: *Modifying Columns*

You can also use the ALTER command to modify existing columns within a table. The basic syntax for column modification is below.

```
ALTER TABLE <table_name> MODIFY <column_name> <column_type>;
```

## 3: *Renaming Columns*

You can also use the ALTER command too rename an existing column within a table. The syntax we use to rename a column within a table is:

```
 ALTER   TABLE   <table_name>   RENAME   COLUMN <prev_name> TO <new_name>;
```

## 4: Deleting Tables

To delete a column using the ALTER command, use the following syntax.

```
ALTER TABLE <table_name> DROP COLUMN <target_column>;
```

**NOTE:** As usual, you can use the command ALTER for other purposes or it may have a different syntax based on the engine used.

# Section 9: SQL SELECT Query

*"Most good programmers do programming not because they expect to get paid or get adulation by the public, but because it is fun to program."*

*Linus Torvalds*

The SQL SELECT query or statement is one of the most commonly used commands. We use it to request and retrieve data from a table within the specified database. The select command returns the requested data information of a table known as result-sets. The syntax for the SELECT command is:

```
SELECT    <expressions>FROM    <table>    WHERE <conditions>;
```

In the above syntax, we use the expression parameter to represent the columns we want to select while we use table to specify the target table where we want to pull records from; we can simply this command as:

```
SELECT   <column1,   column2,   columnN>   FROM
<table_name>;
```

To select the entire columns with the table (you may have seen this above), we use the * wildcard.

```
SELECT * FROM <table_name>
```

Let us select certain columns in the film table of the dvdrental database we created earlier.

In pgAdmin, right click the DVDRental database and launch the query tool. (You can also use the command-line). Enter the command

```
SELECT  title,release_year,rating,length  FROM
film;
```

*SQL For Beginners*

```
DVDRental=# SELECT title,release_year,rating,length FROM film;
       title         | release_year | rating | length
---------------------+--------------+--------+--------
 Chamber Italian     |         2006 | NC-17  |    117
 Grosse Wonderful    |         2006 | R      |     49
 Airport Pollock     |         2006 | R      |     54
 Bright Encounters   |         2006 | PG-13  |     73
 Academy Dinosaur    |         2006 | PG     |     86
 Ace Goldfinger      |         2006 | G      |     48
 Adaptation Holes    |         2006 | NC-17  |     50
 Affair Prejudice    |         2006 | G      |    117
 African Egg         |         2006 | G      |    130
 Agent Truman        |         2006 | PG     |    169
 Airplane Sierra     |         2006 | PG-13  |     62
 Alabama Devil       |         2006 | PG-13  |    114
 Aladdin Calendar    |         2006 | NC-17  |     63
 Alamo Videotape     |         2006 | G      |    126
 Alaska Phantom      |         2006 | PG     |    136
 Date Speed          |         2006 | R      |    104
 Ali Forever         |         2006 | PG     |    150
 Alice Fantasia      |         2006 | NC-17  |     94
 Alien Center        |         2006 | NC-17  |     46
 Alley Evolution     |         2006 | NC-17  |    180
 Alone Trip          |         2006 | R      |     82
 Alter Victory       |         2006 | PG-13  |     57
 Amadeus Holy        |         2006 | PG     |    113
```

If you are using the same database, you will get the same result whether using pgAdmin or the command-line.

## Optional Clause

The SQL SELECT statement supports various parameters. They include:

- ❖ WHERE – We use this parameter to specify which rows to select

- ❖ GROUP BY – We use this parameter to group rows sharing a specific property in order to apply an aggregate function to the specific group.

- ❖ ORDER BY – We use this parameter to set the order in which rows are returned.

## SQL Select Distinct

We use the SQL SELECT DISTINCT command to retrieve only distinct or unique data by the use of a keyword. We use this syntax in cases where a table contains duplicate values and we only require the unique values. The general syntax for the SELECT UNIQUE is:

```
SELECT    DISTINCT    <colum1,   column2,
columnN>FROM   <table_name>;
```

Let us select the distinct values in the rating column of the film table in the DVDRental database.

SELECT DISTINCT rating FROM film:

```
DVDRental=# SELECT DISTINCT rating FROM film;
 rating
--------
 R
 G
 NC-17
 PG-13
 PG
(5 rows)

DVDRental=#
```

This helps us determine that within a movie, there exist only 5 types of ratings.

## SQL Select Unique

The SQL UNIQUE command is similar to the DISTINCT command. Its primary use was in Oracle databases and was declared deprecated. However, some old database models still use it and it is therefore worth noting just in case you come across databases that use it.

## SQL Select Count

The SQL Count is a function used to return the number of rows in an SQL Query. If you have a database with a large amount of data, you can use the count() function together with the SELECT command. For example, in the DVDRental database, we can count the number of customers using this command. The general syntax is:

```
SELECT COUNT (column) FROM table;
```

```
DVDRental=# SELECT COUNT(customer_id) FROM customer;
 count
-------
   599
(1 row)

DVDRental=#
```

We can see from the above output that we have 599 customers. We can also use the wildcard symbol to get the number of records in the specified table.

```
DVDRental=# SELECT COUNT(*) FROM film;
 count
-------
  1000
(1 row)

DVDRental=#
```

From the above Query, we can see that there are 1000 records within the film table. We can use the Count function in conjuction with other commands such as DISTINCT.

## SQL Select Top

We use the SQL TOP command to select data based on the specified number of rows. It helps to specify how many rows from the top should be returned.

The basic syntax is:

```
SELECT    TOP(number_of_rows)    *    FROM <table_name>;
```

The syntax for getting the first elements in various database engines is different. MySQL and PostgreSQL use almost similar syntaxes. The following illustration shows the top 100 rows in the film table of the DVDRental database.

## SQL Select Last

The SQL Last is a function used to fetch the value of the column defined. The general syntax for the last function is:

```
SELECT LAST <column_name> FROM <table_name>;
```

However, and as we have noted many times, syntaxes differ based on the engine.

Below is the syntax for PostgreSQL and Oracle respectively.

### *For PostgreSQL*

```
SELECT <column_name> FROM <table_name> ORDER BY <column_name> DESC LIMIT <limit_number>;
```

### For Oracle Databases

```
SELECT <column_name> FROM <table_name> ORDER BY <column_name> DESC WHERE ROWNUM <=1;
```

```
DVDRental=# SELECT rating FROM film ORDER BY rating DESC LIMIT 2
DVDRental-# ;
 rating
--------
 NC-17
 NC-17
(2 rows)

DVDRental=#
```

# SQL Select First

We use the SQL first function to retrieve the last value in a specified column. It is the opposite of the last function. Other database engines such as PostgreSQL, MySQL, and PostgreSQL do not support this function. Microsoft Access supports it.

```
SELECT FIRST (column_name) FROM <table_name>;
```

## SQL Select Random

We us this function to get random rows within a table in the specified database. We can also use it to fetch random data such as a game. Similar to other SQL functions, the method of fetching random data from a database depends on the engine in use. To fetch random data records or rows in PostgreSQL, we use the following command.

```
SELECT title FROM film ORDER BY RANDOM() LIMIT <number>;
```

There are other complex ways to select random items in PostgreSQL and other engines. For example, to select random rows in MySQL, the general syntax can be as follows:

```
SELECT <column_name> FROM <table_name> ORDER BY RAND ( ) LIMIT <limit_number>;
```

## SQL Select Multiple

We use the SQL multiple command to select table fields from multiple selected tables. We usually accomplish this task by using an SQL join query from the selected tables.

## SQL Select Date

We use this command to retrieve data from a specified database. We can also use it to get database records that match a particular date, before or after the selected date. The date must be correct in the order of YYY-MM-DD.

You can use this function together with other SQL commands such as BETWEEN to get data that matches a specified date.

## SQL Select Null

We use the SQL select Null command to retrieve data where the values are null. As discussed earlier, null values are not spaces or zero. Null values represent missing data. The general syntax for null values data retrieval is:

```
SELECT   <colum1,   column2,   column3…columnN>
FROM<table>WHERE <null_check_column> IS NULL;
```

As mentioned severally, the syntax for null retrieval will depend on the database engine used.

# Section 10: SQL Clauses

*"Gates is the ultimate programming machine. He believes everything can be defined, examined, reduced to essentials, and rearranged into a logical sequence that will achieve a particular goal."*

**Stewart Alsop**

We can classify an SQL clause as any logical SQL statement. We have already discussed some of the SQL clauses such as SELECT, INSERT, BETWEEN and more.

In this section, we are going to expand what we have learned thus far about important SQL clauses such as WHERE, AND, OR, and WITH

## SQL - WHERE Clause

This SQL clause is a conditional clause. We normally use it to set the condition for the required data in a table. We can also use the SQL - WHERE Clause when fetching data from single or joined multiple tables, but its main use is to filter data records and to get the needed records only. We can also use it conjunction with other SQL statements such as SELECT, DELETE, DROP, UPDATE, etc. The basic syntax for the where clause is:

```
SELECT <column1, column2, columnN> FROM <table> WHERE [condition];
```

The WHERE clause accepts conditions such as expressions from operators, both of which we discussed in previous sections.

Let us see an example using the customer table in the DVDRental database.

Using the command:

*SQL For Beginners*

```
select customer_id, first_name, email from customer where address_id > 100;
```

We can be able to fetch all the customer information where the home address is greater than 100. You can also use the WHERE clause to fetch information about a specific customer using that customer's name.

For example, let us try to get information where the first_name is Emma

From the above Query, we can see that the database contains only one customer with the name Emma. This helps us to get only the required information instead of browsing through the entire database scrolling for the name Emma.

## SQL – AND and OR Clause

We normally use the SQL AND & OR clauses to combine two or more conditions. They mainly behave like the AND & OR logical operators. They help minimize the data returned in an SQL statement. We can also consider them conjunctive operators as they help to use multiple SQL operators in a single SQL statement.

Similar to the WHERE clause, we can use this clause with other SQL statements such as DELETE, DROP, UPDATE, etc. Let us see some examples of the AND & OR clauses.

**NOTE:** If you try to perform a delete command against a column referenced by other tables, this will result in an error as shown below.

## SQL – WITH Clause

We use the SQL WITH clause for sub-queries blocks referenceable within the main SQL query. It provides a way to write auxiliary statements for use in larger SQL queries. The syntax for WITH queries may vary based on the engine used. The syntax for WITH in a sub-query alias is:

```
WITH <alias_name> AS (sql_sub-query_statement) SELECT column_list
FROM <alias_name> [table name] [WHERE <join_condition>];
```

The next section of the guidebook discusses how to work with SQL ORDER:

# Section 11: SQL ORDER

*"The computer programmer is a creator of universes for which he alone is the lawgiver. No playwright, no stage director, no emperor, however powerful, has ever exercised such absolute authority to arrange a stage or field of battle and to command such unswervingly dutiful actors or troops."*

*Joseph Weizenbaum*

In this section, we are going to look at SQL ordering statements. These order statements allow us to sort data in a certain order.

## Order By

We use the SQL Order By statements for data sorting. It helps users organize the retrieved data in ascending or descending order. Depending on the database used, sorting of the returned data may be in ascending or descending order.

The general syntax for the Order By query is:

```
SELECT   <column1,   column2…columnN>   FROM
<table> WHERE <condition> ORDER BY <column1,
column2…columnN> [ASC | DESC];
```

The ORDER BY statements accepts any number of columns.

For example:

Let us try to order the data from the film table in the DVDRental Database.

```
SELECT  title,length,last_update  FROM  film
ORDER BY rating, title LIMIT 10;
```

Using the statement: above, we can can be able to order the film titles.

## Order By Ascending

We use the SQL ORDER BY ASC statement to retrieve data from a database and sort it in ascending order. By default, some database engines sort out data in ascending order. However, you can specify the order by during command execution.

The general syntax to include ascending sort in SQL ORDER BY statement is:

```
SELECT    <column-list>    FROM    <table>    WHERE
<specific_condition>           ORDER           BY
<column_sort>ASC;
```

## Order by Descending

The SQL ORDER BY DESC clause is similar to the ASC except that it sorts out data in descending order. Not many engines may sort out data in descending order by default. To specify in which order to sort out the data, you use the syntax above (the ascending syntax) and change the ASC value to DESC.

## Order Data Randomly

If you want to sort out the result of a query randomly, you can use the SQL ORDER BY RAND() function. It is very rare that you will want to sort out data randomly. However, in a case where you want to sort out, say, blog posts, users, etc. randomly, this function will prove very valuable. The general syntax for randomly sorting out data is as:

SELECT <column_list> FROM <table> ORDER BY RAND () LIMIT <number>;

The syntax may vary based on the database engine used. To sort out data randomly in PostgreSQL, the syntax is:

SELECT <column_list> FROM <table> ORDER BY RANDOM () LIMIT <number>;

The case is also different for Oracle SQL Server and IBM DB.

# Section 12: SQL INSERT

*"In theory, there is no difference between theory and practice. But, in practice, there is."*

**Jan L. A. van de Snepscheut**

We use the SQL Insert statement to insert records in table. It can support both single and multiple data input. In SQL, you can insert data in two ways:

❖ By using the SQL Select Insert statement

❖ By using the SQL INSERT INTO statement

We have already covered the select insert statement. However, if you want to use the SQL INSERT INTO query, you can do it in two ways:

❖ You can specify the column name that you want to insert your data. This case is a necessary when inserting partial values in a column.

❖ You can ignore the column names but with a slightly different syntax compared to when you specify the column.

When you have specified the column names, the general syntax is as shown below:

```
INSERT INTO <table_name> [(column1, column2, column3… columnN)] VALUES (value1, value2, value 3, .... Value N);
```

If you do not specify the column names when inserting data, the syntax changes to the following

```
INSERT INTO <table_name> VALUES (value1, value2, value3, ....ValueN);
```

## SQL Insert Multiple

Multiple programming instances will require you as a programmer to insert data into multiple rows. Instances where you can insert multiple rows in a single table of a database using a single SQL statement normally depend on the SQL engine used. SQL Server (2008 version and above) uses the Row Constructor to perform this task. You can learn more about this from the resource page below:

https://bit.ly/2lJLz7G

For PostgreSQL, you can reference the documentation

https://bit.ly/2kn2JrH

We shall now move on to discussing SQL JOIN:

# Section 13: SQL JOIN

*"Learning to program has no more to do with designing interactive software than learning to touch type has to do with writing poetry"*

**Ted Nelson**

The SQL JOIN statement is a very common query used to join tables. It can combine two or more tables in a database by taking the records in the tables and merging them together.

There are five common types of SQL JOINs.

- ❖ Inner Join
- ❖ Right Outer Join
- ❖ Left Outer Join
- ❖ Full Outer Join
- ❖ Cross Join

**NOTE:** The process of SQL JOIN involves combining the rows of the involved tables into one table set.

## SQL Inner Join

The SQL inner Join is the simplest form of SQL join also called EQUIJOIN. It creates a result table by combining the values of the table1 and table2 using the join-predicate logic. This query compares the rows of the first table with the rows of the second table and finds all pairs that match the predicate-join logic.

Consider the following tables:

Customer table

| ID | Customer_name | Customer_age | address | Offered_package |
|---|---|---|---|---|
| 1 | EMMA | 20 | UTAH | 20000 |
| 2 | DANIEL | 22 | OREGON | 22000 |
| 3 | NATASHA | 24 | WASHINGTON | 24000 |

Payment table

| Payment_code | Date | id | address | Offered_package |
|---|---|---|---|---|
| 101 | 9/3/201 | 1 | UTAH | 2000 |

| | 9 | | | |
|---|---|---|---|---|
| **201** | 9/4/2019 | 2 | OREGON | 2200 |
| **301** | 9/5/2019 | 3 | WASHINGTON | 2400 |

If we were to join the above tables using the inner join, we can use the SQL statement shown below.

SELECT ID, Customer_name, Customer_age, Offered_package

FROM Customer c, Payment p

WHERE s.ID =p.Customer_ID;

The above query will combine the customer_id, nam, age, and amount.

## Right Outer Join

We use the Right Outer join to combine the records of the selected tables without considering whether they match or not. The right outer join combines the matching columns on both tables and returns a null value if there is no matching column.

## Left Outer Join

The Left Outer join or left join works in a manner similar to the right join except that it works on rows. It combines the matching rows of both tables and returns a null value if there is no matching row value.

The following shows the syntax for left outer join.

```
SELECT <table1.column1, table2.column2>....FROM <table1> LEFTJOIN <table2> ON <table1.column_field> = <table2.column_field>;
```

## SQL Full Join

The SQL full join is the combination of both right and left join. It works by setting the null value for match not found case for both columns and rows. It is close to the SQL Join.

## Cartesian/Cross Join

Cartesian join occurs when each row of the left table is joint with the rows of the right table. It is mathematically correct to say that it returns the cartesian product of the row sets of the joined table.

# Section 14: SQL KEYS

*"Many people tend to look at programming styles and languages like religions: if you belong to one, you cannot belong to others. But this analogy is another fallacy."*

**Niklaus Wirth**

A database key is an SQL attribute used to identify a tuple in a database table. Database keys allow for the identification of the relation between two or more tables. We also use them to identify (distinctively) rows in a table by grouping of two or more columns in the table. If you set a column in table as the primary key, it automatically adds no duplicate constraint in the column and it cannot contain any duplicates.

As an example:

Consider the DVDRental database we are working with. If we look at the film table, we can see that it contains a primary key as public.film_key. If you navigate the database, you will find that among the constraints within the table is the public.film_key as shown below:

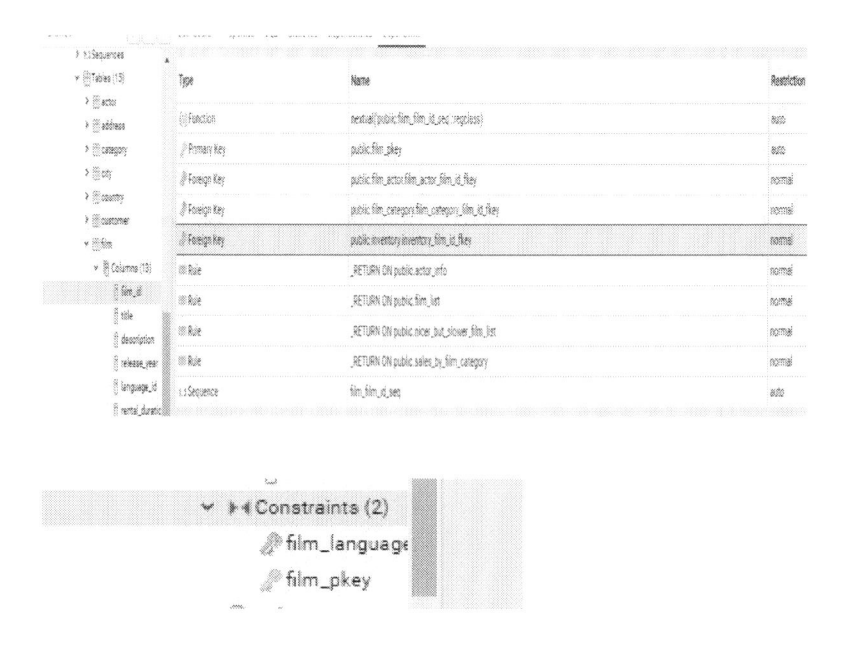

Keys are a very important attribute in database because they help in various ways.

The following are some of the reasons why keys are so important in database management.

* They help to establish the relationship between tables

* They help to identify the relationship between the tables

- ❖ They help reinforce database tables' relationship and integrity

- ❖ They help uniquely identify the database records in case duplicate records.

Databases use various types of keys. Some are more common than others and play and important role once enforced. They include:

- ❖ Primary Keys

- ❖ Super Keys

- ❖ Candidate Keys

- ❖ Alternate keys

- ❖ Foreign Keys

- ❖ Compound Keys

- ❖ Composite Keys

- ❖ Surrogate Keys

Since some keys play similar roles, we shall only discuss the most imporant of these keys.

## Primary Keys

In a database, a primary key is a column used to uniquely identify a row in the same table. As seen above, once you reinforce a key on the column, it cannot contain duplicates, which is to mean there can be no instances of the same value in that table.

### *Primary key rules*

Definition of primary key in a database table follows certain rules. These rules include:

- *The field used to define the primary key MUST not be NULL*
- *The same primary key values cannot be used in more than one row, which is one key for one specific row*
- *Every row must have a primary key value*

- *The value of the primary key CANNOT BE MODIFIED if there are foreign keys that refer to it*

- *A table MUST only comprise of one primary key constraint.*

You can use the following SQL syntax to look for table keys.

```
SELECT GROUP_CONCAT(COLUMN_NAME), TABLE_NAME

FROM INFORMATION_SCHEMA.KEY_COLUMN_USAGE

WHERE

TABLE_SCHEMA = '**dd_name**'

AND CONSTRAINT_NAME='primary'

GROUP BY TABLE_NAME;
```

**NOTE:** The syntax for key check may vary based on the engines. Keys are one of the SQL attributes that vary a lot based on the database engine in use.

## Super Keys

A super key is a collection of one or more keys that uniquely identifies rows in a database table. It usually comprises of more attributes than other keys that may not be essential for exceptional identification of database records.

Suppose you have a table that has person's phone number and social security number. Since the two columns are unique, we can identify them as super keys.

## Candidate Keys

A candidate key mainly refers to a super-key that does not contain redundant attributes. Ensure that every table contains at least one candidate key. A candidate key contains several key properties that include:

- ❖ Must not be redundant – contains only Uniquely identified values
- ❖ Cannot contain a null value

❖ A candidate key may contain more than one attribute – but not similar.

❖ A candidate key should distinctively identify each record in a database table.

| StudentID | Roll Number | FirstName | LastName | Email |
|---|---|---|---|---|
| 147852 | 100 | Peter | Green | greenp@gmail.com |
| 254185 | 200 | Natasha | Dale | dalenat12@outlook.com |
| 54120 | 300 | Hanna | Fox | foxann@ann.org |

From the above table, we can classify StudentID as the primary key. StudentID, Roll Number, and Email are the candidate keys while Roll Number and Email are the alternate keys (discussed next).

## Alternate Keys

Alternate keys consist of candidate keys that are not primary keys. In simple terms, alternate keys are all the other keys that are not primary keys. In the above example, the alternate keys are Email and Roll Number as they are candidate keys BUT not primary keys. It is correct to say that a table can have multiple potential primary keys. For example, in the above table, either Roll Number, Email or StudentID can be set as the primary key as they all contain unique values.

## Foreign Keys

A foreign key is column set to create a relationship between two tables. We normally use foreign keys to preserve data integrity of the database. A foreign key reinforces each relationship in every database model. Foreign keys also help in the triangulation of two or more distinct instances of a single database entity. Consider the following two tables:

| AdminID | FirstName | LastName |
|---|---|---|
| 58452200 | Qwerty | Wesly |
| 58545220 | Janet | World |
| 74575510 | James | Bond |

| CompanyCo | CompanyNa |
|---|---|
| 7845 | TransLogics |
| 7855 | Euro Acres |
| 8440 | TransiNet |

In the tables above, we can see companies and administrators. However, we cannot tell which administrator administers which company. We can create a foreign key between the companyCode and FirstName. This creates a relationship between these two tables where each administrator is matched with their companies.

## Compound Keys

A compound key is one or more field that helps distinctively distinguish specific database records. The case of composite keys in not common; However, a case may rise where a column may not be unique as a single. To make it unique, you can combine the composite keys (discussed next) of other columns to make it unique.

## Composite Keys

A composite key is a key that more than attributes that distinctively identifies rows in a database table. The difference between the composite key and compound key is that composite key can or cannot be a part of foreign key. A compound key on the other hand can contain a foreign key in any part.

## Surrogate Keys

Surrogate keys are keys created in the absence of normal primary keys. Their main intention is to describe each record in a database distinctively. Surrogate keys are mainly set as an integer and do not have any meaning to entities in a table.

# Section 15: SQL Functions

*"Commenting on your code is like cleaning your bathroom – you never want to do it, but it really does create a more pleasant experience for you and your guests."*

**Ryan Campbell**

In programming, a function is a block of organized code that you can reuse to perform a particular operation against a supported data type. SQL has a range of functions that help manipulate numeric and string data.

In this section, we are going to look at some of the most common SQL functions that you can use to perform important aspects of data.

## SQL Numeric Functions

SQL numeric functions, also known as aggregate functions are used to perform mathematical or statistical operations against a specified data type and relations. The list below shows the most common aggregate functions on SQL data.

- ❖ SUM() – used to return the sum of all specified values.

- ❖ MIN() – returns the minimum value within a given set of values

- ❖ MAX() – returns the maximum value within a given set.

- ❖ AVG() – used to get the average of the total values given. It is also referred to as the mean function.

- ❖ COUNT() – used to return the number of instances of a specified value in a given set.

- ❖ SQRT() – used to get the mathematical square root of a given value.

- ❖ RAND / RANDOM() – used to generate a random value or sort values in a random order.

- ❖ LOG() – returns the natural logarithmic value of the given numerical data.

- LOG10() – returns the natural logarithmic value to base 10 of the specified numeric value.

- ABS() – returns the absolute value of the specified numeric value.

- Trigonometric functions – contains a set of function used to perform trigonometric operations of a given value. They include: acos(), asin(), atan(), atan2(), sin(), cos(), tan(), etc.

## SQL String/Text Functions

We use SQL string functions to perform operations on string data types. Their main use is to manipulate string values to return a string or a numerical value. Below is a list of the top SQL string functions:

- UPPER() – used to convert a given set of string/text to uppercase. It can also be represented as UCASE()

- LOWER() – converts the given set of string to lowercase.

- ❖ SUBSTRING() – used to return the substring of given set of string. A substring is selected sequence of characters to form a certain string. For example, a string "database" can have substrings like: "databas", "databa", "datab", data" etc.

- ❖ CONCAT() – used to add two or more strings together to return on set of sting.

- ❖ CONCAT_WS() – used to add strings together using a specified separator.

- ❖ HEX() – returns a string illustration of a hex value.

- ❖ LENGTH() – used to return the value of the specified string in bytes.

- ❖ LOCATE() – used to return the first occurrence of a substring.

- ❖ REPEAT() – used to copy/repeat a set of string using the set number of times.

## SQL Date Functions

Just as there are date data types in SQL, there are specific functions used to operate on this kind of data. Date functions vary significantly based on the database engine used.

Here are the main date functions for PostgreSQL

- AGE(timestamp, timestamp) – used to return the difference between two timestamps as a symbolic value in years and months.

- AGE(timestamp) – returns the difference between timestamp and current_date(midnight).

- CURRENT_TIME() – returns the current time during the function call.

- CURRENT_DATE() – returns the current date during the function call.

- TIMEOFDAY() – returns the current time and date at the time of function call.

❖ ISFINITE() – Used to test for finite date.

# Section 16: SQL Injections

*"Don't worry if it doesn't work right. If everything did, you'd be out of a job."*

***Mosher's Law of Software Engineering***

We use databases in mobile applications, local stores, online stores, social media, and even android applications. If you are using an application that connects to a database, you may have to be careful of how you program the database.

In this section, we are going engage in a brief discussion of a key vulnerability in SQL based applications known as SQL injection.

# What is SQL Injection?

SQL injection refers to a vulnerability in poorly programmed SQL based applications. This vulnerability allows users to add SQL commands directly into the database. They mainly occur when a user input includes illegal characters interpreted as SQL commands in the database. Consider the following code:

```
if (preg_match("/^\w{8,12}$/", $_GET['username'], $matches)) {
    $result = mysql_query("SELECT * FROM CARDNUMBER
        WHERE name = $matches[0]");
} else {
    echo "Incorrect Username";
}
```

You can see that the name is set to only alphanumeric characters between the length between 8 and 12. If we provide the following malicious:

```
$name = "Incorrect'; DELETE FROM CARDNUMBER;";
mysql_query("SELECT * FROM CARDNUMBER WHERE name='{$name}'");
```

The function call from the above will be programmed to get a record from the CARDNUMBER table in the instance of a column name matching the name provided by the user. The above code would work perfectly if only the provided information follows the set rules of alphanumeric characters. However, in the case where a malicious SQL query is appended to the name variable, the entire CARDNUMBER table ends up being deleted.

Cases of SQL Injections are very popular; the only way to eliminate it is to employ strict and precise programming. The following are some factors to consider as a way to prevent malicious SQL query injections from user input.

- Always set a validation code – This ensures that the user input is validated before being submitted for processing.

- Use parameterized queries – these are queries that treat all the user input as input rather than commands.

- Constraint the database – Set constraints to database table to prevent malicious operations; for example, set a password require constraint for deletion, update, and cases of data retrieval from the database.

- Hide sensitive information – Do not display errors to the user. Never display error information to the user. Instead, log all the error information to a file for later evaluation.

In the next section, we shall discuss a step-by-step data manipulation project that helps you get hands-on experience with the various SQL elements we have discussed throughout the last 16 sections of this guide.

# Section 17: A Step-by-Step SQL Data Manipulation Exercise

*"Optimism is an occupational hazard of programming; feedback is the treatment."*

**Kent Beck**

## Exercises & Projects

In this section, we are going to perform exercises and projects that will help sharpen our skills.

### Exercise 1

Using the DVDRental database, retrieve the first 20 film titles that had a rental_duration of less than 6.

### Sample Solution

You can retrieve the film titles using the conditional check with the SQL WHERE statement. The simplest way is:

```
SELECT  t.title  FROM  public.film  t  WHERE
rental_duration < 6 LIMIT 20;
```

**NOTE:** To ensure that you use your own method to retrieve the data, I have used a different syntax.

## Exercise 2.

Retrieve the Customer IDs of the customers who have spent at least 110.

### **Sample Solution**

You can also get all the IDs of the customers who have used at least 110 dollars.

The sample code would be:

```
SELECT              customer_id, SUM(amount)
FROM                             public.payment
GROUP           BY                   customer_id
HAVING SUM(amount) > 110;
```

## Exercise 3

*SQL For Beginners*

Get the total number of films that has a title that start with the later 'A'

## **Sample Solution**

```
SELECT          *          FROM          film
WHERE title LIKE 'A%';
```

# BONUS SECTION! An Interactive SQL Project

In this section, we are going to practice the process of creating a database from scratch and then adding data and primary keys maintaining data integrity.

## Step 1

Create a new database and name it as PractiseDb. Do not use the graphical control (pgAdmin) for this part.

```
create database Practise_Db;
```

## Step 2:

Using one SQL script, create a new database schema and name it practice_schema. Next, using the schema created, add a table, and add a description as follows "a table to store product information"    .

## SQL For Beginners

In the above table, create five different columns as follows: product_id, product_name, department, price, and date.

Set the data type of the above columns as follows (respectively): integer, text, text, money, and date. Set the product_id as product key. The following columns MUST be set to contain no NULL value (meaning cannot be empty): product_id, date, product_name.

Add a relationship (foreign key) from product_id to product_name and set the delete rule to restrict.

**NOTE:** Use the SQL commands and not pgAdmin. A sample code is below:

```
create table practise_store
(
    product_id          integer     not null
        constraint practise_store_pk
            primary key,
    product_name text                not null,
    department          text         not null,
```

```
    price                               money,
    date               date       not null
);


comment on table practise_store is 'a table
to        store         information';

alter      table           practise_store
    owner             to           postgres;

create         unique              index
practise_store_product_id_uindex
    on practise_store (product_id);
```

Once you have created a database, a database schema, tables and columns, launch pgAdmin and enter sample data to the columns. Intentionally experiment with the database using the knowledge you have distilled from this guide; for instance, you can try adding a text to the product_id column.

# **Conclusion**

*"There are two ways of constructing a software design: One way is to make it so simple that there are obviously no deficiencies, and the other way is to make it so complicated that there are no obvious deficiencies. The first method is far more difficult."*

**C.A.R. Hoare**

As is the case with most programming languages, to master SQL, what you really need to do is to learn by practicing. Use the knowledge you have learned to interact with the database you just created in this project and practice working within the SQL environments we created in [section 3 of this guidebook](#).

If you are happy with what you've learned, don't forget to leave a review of this book on Amazon.

## Check Out My Other Books

**Kali Linux: Kali Linux Made Easy For Beginners And Intermediates Step By Step With Hands On Projects (Including Hacking and Cybersecurity Basics with Kali Linux)**

**JavaScript: JavaScript Programming Made Easy for Beginners & Intermediates (Step By Step With Hands On Projects)**

Printed in Great Britain
by Amazon